# EARTHQUAKE SURVIVAL GUIDE

## PREPARING FOR THE UNEXPECTED

**BEST SELLER**

"The Ground Beneath Your Feet Holds Immense Power – Are You Truly Prepared? When the Earth Trembles, Every Second Counts – This Guide Could Save Your Life."

### Alex Anderson

# EARTHQUAKE SURVIVAL GUIDE PREPARING FOR THE UNEXPECTED

"A Comprehensive Guide to Understanding and Preparing for Earthquakes"

BY

ALEX ANDERSON

# Introduction

**"Earthquake Survival Guide: Preparing for the Unexpected"** is a book that tries to educate readers on how to be ready for an earthquake on a personal, family, and community level. Recognize the different types, causes, and measurement scales of earthquakes; places that are susceptible to seismic activity; create emergency plans; respond to safety concerns; evaluate damage; handle injuries; seek cover; deal with post-earthquake shock; and comprehend the psychology and feelings associated with experiencing such events. The need of safeguarding irreplaceable papers and belongings, animals and pets, and individuals with special needs or impairments is also covered in the book. This book aims to provide readers with the knowledge and tools necessary to navigate the challenges posed by natural catastrophes while maintaining their safety and fortitude.

# Dedication

This dedication is for readers, scientists, survivors, heroes, and anybody who has felt the force of Mother Nature. It gives gratitude and praises those who have persevered and come out uninjured. The book also acts as a monument to the human spirit, inspiring readers to meet adversity head-on with bravery and resilience and demonstrating the enduring strength and resilience of the human spirit.

# **Disclaimer**

This book provides a general guide for earthquake preparedness, response, and recovery, based on research and recommendations from reputable sources. It is not responsible for errors or misuse of the information. The strategies are meant to serve as guidelines and should be adapted to specific situations. Readers are encouraged to exercise caution and seek professional advice when necessary. The author and publisher are not liable for any injury, damage, or loss resulting from the use of the information. The responsibility for personal safety and preparedness lies with the reader, and they should approach the information with appropriate expectations and responsibility.

Copyright © 2024 by [Alex Anderson]

ISBN: 9798327101531

All rights reserved. No part of this book may be reproduced, stored in a retrieval system, or transmitted in any form or by any means, electronic, mechanical, photocopying, recording, or otherwise, without the prior written permission of the publisher, except for brief quotations in reviews and certain non-commercial uses permitted by copyright law.

This book is a work of nonfiction. Unless otherwise noted, the author and publisher make no explicit guarantees as to the accuracy of the information contained in this book and disclaim liability for damages or inaccuracies arising from the use of the information within. This book is intended as a resource to aid in earthquake preparedness but should be used in conjunction with other authoritative sources and professional advice.

# TABLE OF CONTENTS

**I. Introduction** ------------------------------------ **pg3**
   A. The importance of earthquake preparedness
   B. Statistics and facts about earthquakes and their impact
   C. Overview of the book's content and purpose
II Dedication ------------------------------------pg4
III. Disclaimer ----------------------------------pg5
IV. Copyright claim -----------------------------**pg6**

**Chapter 1:Understanding Earthquakes**----------**pg9**
   A. What are earthquakes?
   B. Causes and types of earthquakes
   C. Earthquake measurement scales (magnitude, intensity)
   D. Areas prone to earthquakes

**Chapter 2:Preparing Your Home**------------------**pg15**
   A. Securing furniture and appliances
   B. Creating a safe room or space
   C. Emergency supplies and kits
   D. Storing food and water
   E. Securing utilities (gas, electricity, water)

**Chapter 3:Developing an Emergency Plan**--------**pg19**
   A. Creating a family emergency plan
   B. Establishing communication channels
   C. Designating meeting points
   D. Preparing emergency kits for home, work, and vehicles

**Chapter 4: During an Earthquake――――――――pg23**
   A. Drop, cover, and hold on
   B. Safety measures inside buildings
   C. Safety measures outside buildings
   D. Responding to specific situations (e.g., in a car, at work, at school)

**Chapter 5: After an Earthquake―――――――――pg 29**
   A. Assessing damage and safety
   B. Handling injuries and first aid
   C. Finding shelter and resources
   D. Dealing with aftershocks

**Chapter 6: Special Considerations――――――――pg 31**
   A. Preparedness for people with disabilities or special needs
   B. Preparing pets and livestock
   C. Protecting valuable documents and belongings

**Chapter 7: Recovery and Resilience――――――――pg36**
   A. Emotional and psychological impact
   B. Rebuilding and restoring your home
   C. Accessing disaster assistance and resources
   D. Learning from the experience

**Conclusion―――――――――――――――――pg42**
   A. Recap of key points
   B. Importance of staying informed and prepared
   C. Additional resources and information sources

# Chapter 1
# Understanding Earthquakes

## What are earthquakes?

Earthquakes are sudden and violent shocks from the earth's surface, caused by the release of energy from the Earth's crust. This energy is generated by the constant motion of the tectonic plates, huge rocks that form the Earth's surface. When these

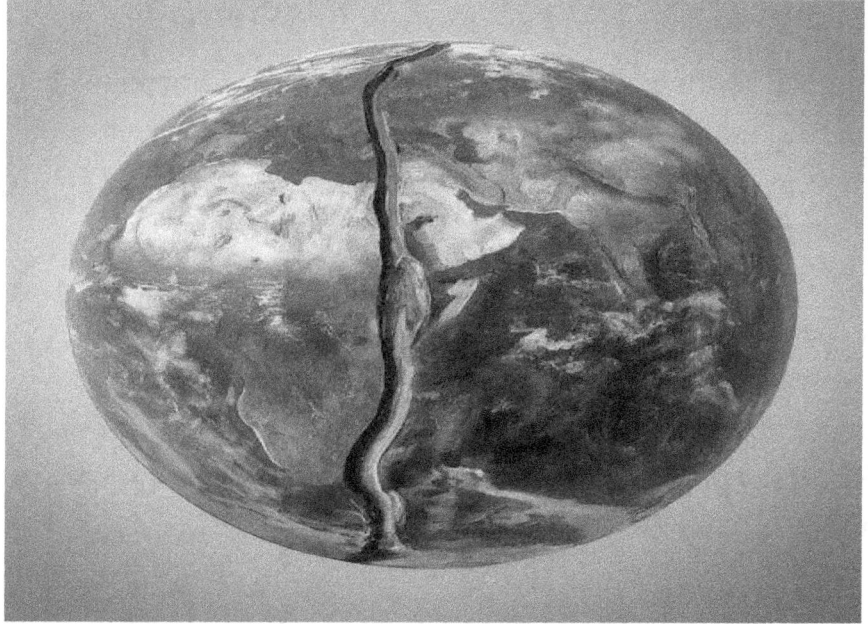

plates move, fry, or collide with each other, the resulting friction and pressure increase can cause the crust to break, releasing enormous amounts of energy in the form of seismic waves.

# Causes and types of earthquakes

## Movements of tectonic plates
The Earth's crust consists of several important tectonic plates, including the Pacific, North American, Eurasian, and Antarctic plates among others. These plates are in constant motion, driven by convection currents inside the Earth's mantle. As these plates interact, they can collide, separate, or slip over each other, resulting in different types of earthquakes.

## Failure lines

Earthquakes often occur along faults, which are fractures or ruptures in the terrestrial crust where tectonic plates are located.

### There are three main types of failures:
1. **Slide failures:** occur when two plates slide horizontally over each other.
2. **Normal failures:** occur when one plate moves down relative to the other.
3. **Reverse (push) failures:** occur when one plate is pushed up and over the other.

## Other causes

While most earthquakes are caused by motion of tectonic plates, there are other less common causes, such as volcanic activity, man-made explosions, and the collapse of underground mines or caves.

# Earthquake measurement scales

**Richter Magnitude Scale**

The Richter magnitude scale is a logarithmic scale that is used to measure the strength or magnitude of an earthquake. Each increase of an integer on the scale represents a ten-fold increase in the amplitude of the seismic waves generated by the earthquake. For example, an earthquake of magnitude 7.0 is ten times stronger than a 6.0.

**Modified Mercalli intensity scale**
The Modified Mercalli Intensity Scale measures the intensity or severity of an earthquake in a specific location. It is a qualitative scale ranging from I (no sense) to XII (total destruction). This scale is based on observable effects, such as damage to structures and changes in the natural environment.

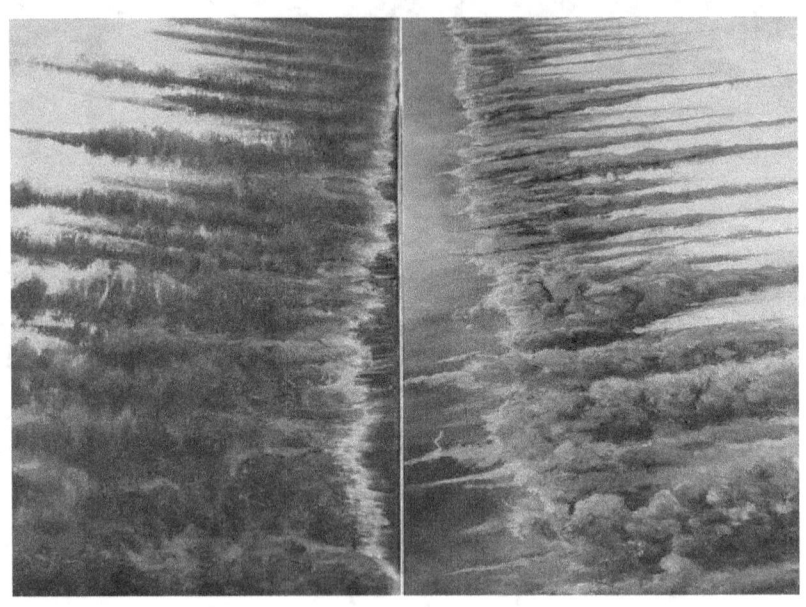

## Earthquake-prone areas

Certain regions of the world are more prone to earthquakes due to their proximity to the borders of the tectonic plates and to the active faults. Some of the areas most prone to earthquakes include the Pacific Ring of Fire, the St. Andrew Falls in California, the Alpine Belt in Europe and Asia, and the Himalayan region.

By understanding the fundamental concepts of earthquakes, their causes, scales of measurement and areas prone to seismic activity, readers will be able to better appreciate the importance of preparedness and the potential impact of these natural disasters.

# Chapter 2
# Getting your house ready

While your home is a place of refuge during an earthquake, it can also pose a serious threat if you are not properly prepared. Uninsured furniture, appliances, and other household items can become deadly missiles, causing injury or damage. This chapter will guide you through key steps you can take to strengthen your home and make it a safe place for you and your family.

**Safely store appliances and furnishings.**
Securing big, heavy furniture and appliances to keep them from sliding or falling during the tremors is one of the most important parts of preparing your house for an earthquake. **The following are some suggested tactics:**

**Bookstores and shelves:** Use flexible straps or corner supports to fasten high shelves and shelters to the wall. Verify that there aren't too many things on the shelves, and think about putting up barriers or lip guards to stop anything from dropping.

**Electronics and televisions:** Computer monitors and flat-screen TVs should be placed on low, robust furniture or securely mounted on the wall. If you want to keep children from sliding or falling, use seat belts or supports.

**Refrigerators and other home appliances:** Use flexible belts or supports made specifically for this purpose to secure refrigerators, ovens, and other household appliances to the wall or floor. Additionally, to reduce the possibility of gas leaks during an earthquake, make sure the appliances that use gas have flexible couplings.

**Establish a secure area or room.**
Having a designated safe room or area inside your house might give an extra degree of security in the event of an earthquake. This space should be in the middle of the home, away from any windows and potential trip hazards. Think about adding strong furnishings or materials that may provide more protection to this area in order to reinforce it.

**kits and supplies for emergencies**
After an earthquake, it's critical to have the necessities on hand to ensure your safety and wellbeing. Think about assembling an emergency supply bag with the following items:

- **First Aid Package**
- **Extra batteries and flashlights.**
- **Radio on a battery**
- **Water and non-perishable food (enough to last at least three days)**
- **Abrasive materials by hand**
- **An abundance of coats.**
- **Items for personal hygiene**
- **Prescription copies and medications.**
- **Money and essential papers (in an impervious wallet)**
- **Instruments (like a key to cut off public utilities)**

Maintain a readily accessible location for your emergency kit and periodically check and update its contents as required.

## Storage of food and water

Water and fresh food may become scarce after an earthquake. It is advised to have enough non-perishable food and water in stock for each member of your family to last at least three days at home.

Take into account these suggestions:

- **Replace the food-safe, hygienic containers holding the water once every six months.**
- **Stock up on nonperishable goods including energy bars, dried fruits, preserves, and dried fruits.**
- **Verify that you have utensils and a manual abrelatas.**
- **To keep your food supply from running out, break them often.**

**Public service security**

Public utility lines may be damaged by earthquakes, which might lead to water contamination, gas leakage, or electrical fires. Use these safety measures to reduce these risks:

**Gas:** Find your gas meter and learn how to turn it off. Take into account adding an automated gas shutoff valve.

**Electricity:** Acquire the knowledge of securely shutting off the main disconnector or fuse box.

**Water:** Locate the primary water closing valve and have the proper equipment handy to shut it as needed.

You may greatly lower your home's risk of earthquake-related injuries and damage by putting these doable precautions into place. To ensure your safety and the protection of those you love, always remember that preparedness is the key.

# Chapter 3
# Developing Emergency Plans and Procedures

A thorough emergency plan is necessary to ensure a coordinated and effective response during an earthquake. This chapter walks you through the steps to create a basic emergency kit that can be used in any scenario and to develop a comprehensive family emergency plan.

## Developing a Family Backup Plan

A family emergency plan is a roadmap for your family that describes specific actions and guidelines to follow during and after an earthquake.

Here are important factors to consider:

**1. Create a communication channel**: Designate an external contact that all family members can text or call to update their whereabouts and status. Once information is coordinated, the individual can help reunite the family if the family has been dispersed.

**2. Choose a meeting point:** If you are unable to return home or must flee the area, choose two meeting points, one of which should be close to your residence.

**3. Assign roles and responsibilities**: Each family member is given certain tasks, such as closing common buildings, allowing access to emergency supplies, or providing assistance in an emergency.

**4. Practice and review:** To ensure everyone knows their roles and responsibilities, practice the emergency plan with your family regularly. Review and revise as needed.

**Establish a communication channel**
Effective communication is essential during an emergency crisis. In addition to delineating external relationships, consider the following strategies:

**1. Develop a communication strategy for your family**, including key phone numbers and email addresses.

**2. To ensure you receive timely information and instructions,** sign up for your local government's emergency alert system.

**3. To avoid disrupting established routes, discuss alternative** forms of communication, such as texting or using social media.

**Decide on a meeting place.**
Creating a meeting place before an earthquake is essential to avoid family scattering.
 Choose two locations that are well known and easily accessible to everyone:

**1. A meeting place close to your home:** This could be a community center, school, or park.
A meeting place outside your community:
Choose a location far enough away, such as a friend's house or a recognized public place.

**Prepare emergency supplies**
In addition to the home emergency kit (car) described in Chapter 2, you must also prepare emergency kits in other places, such as your workplace or car. These kits need to include survival essentials and the most basic emergencies, **such as:**

**Water and food that will not corrode**
**Emergency supplies:** spare batteries and light bulbs.
**Whistle:** emergency call signal
**Dust emergency blanket and dust mask.**
 **Swiss Army knife or multi-tool**
**Copies of important records and emergency contacts**

By developing a comprehensive emergency plan and gathering the equipment you need, you can improve your chances of safely reuniting with your family and overcoming the hardships of an earthquake.

# Chapter 4
# When an Earthquake Occurs

Amid an earthquake, every moment matters. It may be the difference between life and death to know how to respond swiftly and securely. This chapter will provide brief, understandable guidelines on how to be safe during an earthquake for both you and other people.

**Lower, Shield, and Cling**
***"Drop, Cover, and Hold On"*** is the most crucial tactic to know during an earthquake. This easy-to-use but powerful technique may significantly lower the chance of being hurt by debris and falling items.

1. Quickly collapse to the ground (avoid running outdoors as there may be falling debris).
2. If at all possible, seek shelter behind a strong table or desk and cover your head and neck with one arm. If there's no heavy furniture nearby, kneel against an interior wall and shield your head and neck with both arms.
3. Remain in the shelter until the shaking stops by holding onto it. Use both arms to support your head and neck if you are not seated under heavy furniture.

## Safety Procedures Inside Structures

When inside a structure during an earthquake, there are other crucial safety precautions to take in addition to the "***Drop, Cover, Hold On***" method:

1. Avoid areas with windows, outside walls, and loose, breakable things.
2. Avoid using elevators since they might break down or get trapped.
3. If you find yourself in a high-rise structure, wait to leave until you are told to. Usually, falling debris from outside th e structure poses the most risk.
4. Refrain from hurrying to leave a crowded public area as this may trigger a potentially fatal stampede.

## Safety Procedures for Exterior Buildings

When an earthquake occurs, if you're outdoors, heed these instructions:

1. Keep clear of structures that have the potential to collapse or dump debris, such as power poles and buildings.
2. Avoid electricity wires since they have the potential to fall or get electrified.
3. After you're in a clear space, kneel down and shield your head and neck with your arms.
4. Remain outside until the shaking stops, and keep an eye out for any possible aftershocks.

### Reacting to Certain Circumstances

Since earthquakes may occur anywhere and at any time, it's critical to know how to respond in certain scenarios, such as:

**In a car:** Park in a wide space. Steer clear of power wires, overpasses, bridges, and other obstacles. Put on your seatbelt and remain in your car until the shaking stops.

Observe established emergency planning and evacuation protocols at work or school. If you don't have a plan, employ the *"Drop, Cover, and Hold On"* method and leave the building via the stairs rather than the elevator as soon as the shaking stops.

Avoid rushing to flee in crowded public areas since this may lead to a hazardous stampede. Observe the *"Drop, Cover, and Hold On"* method as you await orders from authorities.

You may considerably improve your chances of surviving an earthquake and reducing your injuries by paying attention to these safety precautions and advice. Recall that it is important to maintain composure and take prompt, decisive action in these potentially fatal circumstances.

# Chapter 5
# After an Earthquake

Healing after an earthquake can be as challenging as the earthquake itself. Once the shaking stops, it is important to take immediate action to ensure your safety and begin the recovery process.

This chapter guides you through the basic steps after an earthquake, from assessing damage and treating injuries to finding shelter and resources.

Assessing Damage and Safety Before attempting to move or taking any other action, it is critical to assess your surroundings for damage and safety.

**Follow these guidelines:**

**1. Check for injuries:** Treat any injuries, whether to yourself or those around you. Provide emergency assistance or seek medical help if necessary, if trained.

**2. Assess your surroundings**: Search for potential hazards, such as damaged buildings, downed power lines, gas leaks, or fires. If you suspect a gas leak, evacuate the area immediately and do not use any fire or electrical equipment.

**3. Check your house:** If you are at home, check for structural damage, such as cracks in the walls, fallen debris, or sloping ground. If the building appears unsafe, evacuate immediately and do not re-enter without a professional inspection.

**4. Turn off utilities:** If you smell gas or suspect a leak, turn off the gas supply at the main valve. In addition, consider turning off electricity and water if these systems are damaged. After an earthquake, injuries range from minor cuts and scrapes to more serious wounds or trauma.

Knowing basic first aid techniques is important to provide immediate care before professional medical help arrives.

**1. Establish a differential diagnosis system:** If multiple people are injured, prioritize treatment based on the severity of the injury, and pay immediate attention to life-threatening situations.

**2. Control bleeding**: Apply direct pressure to any wound or use a tourniquet if necessary to stop severe bleeding.

**3. Immobilize fractures:** Immobilize fractures with pads or other materials to prevent further injury.

**4. Shock monitoring**: Watch for signs of shock, such as pale, cold, wet skin, rapid breathing, and confusion. If signs of shock appear, keep warm, elevate your legs, and lie down.

**5. Seek professional medical help:** Call emergency services as soon as possible or take the injured to the nearest medical facility.

Find shelter and resources. Depending on the extent of the damage, you may need to find temporary shelter and obtain food, water, and other necessities.

**Consider the following options:**
**1. Stay or leave:** If the residence is safe and habitable, it is best to shelter in place. However, if your residence is damaged or unsafe, you may need to evacuate and find alternative shelter.

**2. Emergency Shelters:** Local authorities may have established emergency shelters in schools, community centers, or other public buildings. If directed to a shelter, follow the instructions and follow official evacuation routes.

**3. Resources and Assistance:** Learn about local resources and assistance programs that can provide food, water, medical care, and other basic services after an earthquake.

Remnant earthquakes are small earthquakes that can occur in the days, weeks, or even months after a major earthquake. While remnant earthquakes are typically less intense than the main earthquake, they can cause additional damage and pose a safety threat.

During an earthquake, be prepared, follow the "*squeeze, cover, and hold" technique, and avoid entering or* remaining in damaged buildings until the building has been inspected and determined to be safe.

By following these guidelines and taking action immediately after an earthquake, you can increase your chances of staying safe, getting proper medical care, and accessing the resources you need during the recovery process.

# Chapter 6
# Special Considerations

Though there are general guidelines for earthquake preparedness, certain groups and circumstances call for special thought and preparation. This chapter covers preparation steps for pet owners and livestock management as well as the particular requirements of those with impairments or special needs.

***Special requirements or disabled persons'*** preparations
Special needs or disabled people might have more difficulties both during and after an earthquake.
Emergency supplies and plans must be modified to fit your particular needs. These significant factors should be taken into account:

**1. Mobility aids:** Check that a wheelchair, walker, or other mobility device is on hand and in excellent operating condition. For power-assisted equipment, think about having a battery backup or backup device. .

**2. Communication needs**: Give those with speech or hearing impairments other ways to communicate, like textual materials, pictograms, or text-to-speech devices.

**3. Medical needs**: Maintain a current inventory of prescription drugs, medical supplies, and equipment, and arrange to get replacements or refills as needed.

**4. Service Animal:** Make plans for the feeding, watering, and extra supply preparation and housing of the service animal.

**5. Evacuation support:** Name reliable individuals who may assist with the evacuation or provide aid in an emergency. Talk to them about and become use to evacuation protocols.

**6. Emergency Contacts:** Keep track of all of the people you should call in case of emergency, including medical professionals, family members, and support groups.

**Get the animals and pets ready. .**

Your emergency preparations should give pets' and livestock's safety and well-being first priority since they are equally susceptible during earthquakes. Think about the following actions:

## **Pet**:
1. Fill a food, water, medicine, and other supply emergency bag for your pet.
2. In the unlikely event that your pet becomes lost, make sure it is correctly identifiable (microchip, collar, and tag).
3. If you must leave your house, choose pet-friendly lodgings or shelters.

4. If an evacuation is required, arrange for your pet to be moved and housed. .

## Livestock:

1. During an earthquake, shield barns, stables, and other structures to avoid injury or escapes.
2. Make that the cattle have enough food, water, and medical supplies.
3. Determine backup pastures or food supplies in case storage facilities or pastures are destroyed.

*If needed, create a strategy to transfer and relocate cattle.*

## Save important papers and stuff.

Protecting critical papers and priceless items is just as crucial as ensuring the safety of humans and animals.
Think about the following actions:

**1. Keep vital paperwork (passports, insurance policies, birth certificates, etc.) in a safe or other waterproof, fireproof container.**
**2. Digitally backup critical data, then safely store them on an external hard drive or the internet.**
**3. Recognize and safeguard priceless collections, pictures, or family treasures.**
**4. To aid with insurance claims or, if needed, replacements, make an inventory of your possessions, including pictures or videos.**

Your most valuable belongings will be safe in the case of an earthquake catastrophe and no one will be left behind if you take care of these particular concerns and include them into your overall earthquake preparation strategy.

# Chapter 7
# Recovery and Resilience

The aftermath of an earthquake can be emotionally and psychologically challenging, even for those who have not experienced direct physical harm or property damage. This chapter explores the potential emotional impact of such events and offers strategies for coping, recovery, and fostering resilience within individuals and communities.

**Emotional and Psychological Impact**
Experiencing a traumatic event like an earthquake can trigger a range of emotional and psychological responses, including:

**1. Anxiety and fear:** Feelings of anxiety, fear, and panic are common reactions, especially in the immediate aftermath of an earthquake and during aftershocks.

**2. Grief and loss:** If the earthquake resulted in the loss of loved ones, homes, or cherished possessions, individuals may experience profound grief and a sense of loss.

**3. Stress and overwhelm**: The chaos, uncertainty, and disruption caused by an earthquake can lead to feelings of stress, overwhelm, and difficulty coping with daily tasks.

**4. Post-traumatic stress disorder (PTSD):** In some cases, individuals may develop PTSD, characterized by intrusive thoughts, nightmares, avoidance behaviors, and hypervigilance.

It's important to recognize these emotional responses as normal reactions to an abnormal and traumatic event. Seeking support and professional help when needed is crucial for emotional healing and recovery.

## Reaction and self-care techniques

Self-care and the use of healthy response techniques may assist individuals in managing the psychological and emotional difficulties related to earthquake rehabilitation.
Think about the following strategies:

**1. Seek help**: Share your experiences and emotions with friends, family, or support groups. Expressing your feelings to others may be a very effective coping strategy.

**2. Use relaxation techniques:** Take part in activities like yoga, meditation, or deep breathing exercises that help people relax and relieve tension.

**3. Upholding routines**: Try to create and maintain family rituals and activities as much as you can, as they may provide a feeling of security and normalcy.

**4. Self-care**: Getting enough sleep, maintaining a healthy diet, and engaging in regular physical activity are all important for maintaining good health.

**5. Limit your exposure to traumatic reminders**: Steer clear of news articles or media coverage that might overwhelm you or bring back unpleasant memories or feelings.

**6. Seek professional assistance:** Get therapy or guidance right away if you or someone you love is experiencing severe anxiety, depression, or post-traumatic stress disorder.

# Rebuild and renovate your home

For individuals whose lives were lost or severely damaged after the earthquake, the repair and rehabilitation process may be challenging. Observations and actions to be done include the following:

**1. Records of damage**: images and videos showing the harm done to your home and belongings for insurance reasons.

**2. Collaborating with experts**: providing guidance to contractors, structural engineers, and other experts in determining the degree of damage and creating plans for rebuilding or repair.

**3. Managing insurance claims**: To file a claim and comprehend the terms and conditions of your policy, collaborate closely with your insurance company.

**4. Look into financial aid:** Research and apply for any government or nonprofit assistance programs that may be able to help with the expense of temporary housing or rebuilding.

**5. Safety Priority**: Throughout the rebuilding process, safety is prioritized by abiding by all building standards and rules and taking care of any possible risks or structural flaws.

**6. Reconstruction recovery capacity**: To make homes more earthquake-resistant in the future, take into account the use of seismic design and building methods.

# Obtaining resources and aid in times of catastrophe

Numerous governmental bodies, nonprofits, and neighborhood associations may provide support and resources to help get things back to normal after a significant earthquake. Look seek and use these resources proactively; these might include:

1. **State and federal programs for disaster assistance**
2. **Resources for temporary accommodation and shelter**
3. **Distribution centers for food and water**
4. **Mental health and counseling services**
5. **Low-interest loans and financial aid programs.**
6. **Community support networks and volunteer groups.**

# Take experience to heart.

Although surviving an earthquake may be painful, it can also provide chances for improvement in preparedness, community building, and personal development.

Consider the things you've learnt and use them to enhance your backup plans and recuperation techniques.

To encourage community preparation and increase awareness, share your experiences with others.

By addressing the emotional and psychological issues that arise during the earthquake recovery process, adopting preventative reconstruction and recovery measures, utilizing available resources and assistance, and drawing lessons from past mistakes, people and communities can become stronger and more resilient in the face of future natural disasters.

# IN CONCLUSION

Earthquakes are powerful reminders that natural forces are unpredictable and that it is important to be prepared for the unexpected. In this book, we explore the basics of earthquakes, what causes them, and the areas most susceptible to them. We delve into practical strategies for preparing your home, developing a comprehensive emergency plan, and responding safely during and after an earthquake.

We also address special considerations for people with disabilities or special needs, as well as preparedness measures for pet owners and livestock managers. In addition, we discuss the emotional and psychological impacts of earthquakes and provide guidance for recovery, reconstruction, and building personal and community resilience.

By implementing the recommendations and actionable steps outlined in these pages, you can significantly increase your chances of survival and minimize the likelihood of injury, damage, and loss when an earthquake strikes. Remember that preparedness is not a one-time effort; it is an ongoing process that requires regular review, updating, and practice.

Earthquakes are powerful reminders that it is important to be prepared for the unexpected. While we cannot prevent these natural disasters from happening, we can take proactive steps to mitigate their effects and increase our chances of survival and recovery.

Stay up to date with the latest earthquake preparedness guides and resources for your local community. Work with emergency management organizations, attend preparedness workshops, and participate in drills and exercises to stay prepared.

In addition, share the knowledge and strategies you gain from this book with your family, friends, and community members. Inspire others to take action, prioritize earthquake preparedness, and foster a culture of resilience and collective responsibility.

Remember, preparedness is about more than protecting yourself; it's about protecting the well-being of your loved ones and contributing to the resilience of your community. By working together and taking proactive steps, we can face the unpredictable forces of nature with confidence, resilience, and a shared commitment to survival.

**Stay safe, be prepared, and embrace the valuable lessons and experiences that come with earthquake preparedness.**

# Final Note: Embracing Resilience, Inspiring Preparedness

As we reach the end of this journey through the pages of the **"Earthquake Survival Guide,"** it is important to reflect on the knowledge and insights we have gained. The very act of reading this book is a testament to your commitment to preparedness and a desire to protect yourself and your loved ones from the unpredictable forces of nature.

Earthquakes are a humbling reminder of the immense power that lies beneath our feet, a power that can shake the foundations of our lives in an instant. Yet, it is in these moments of adversity that the indomitable spirit of humanity truly shines. Time and again, we have witnessed communities rise from the rubble, rebuild, and emerge stronger than before.

This book has armed you with the tools and strategies to face the unexpected with courage and resilience. From understanding the science behind earthquakes to fortifying your home and developing comprehensive emergency plans, you now possess a wealth of knowledge to protect what matters most.

But preparedness is not a destination; it is a continuous journey. As you embark on this path, remember to stay vigilant, review and update your plans regularly, and share your knowledge with others. Inspire your family, friends, and community to embrace a culture of preparedness, for it is through collective action that we can truly weather the storms of life.

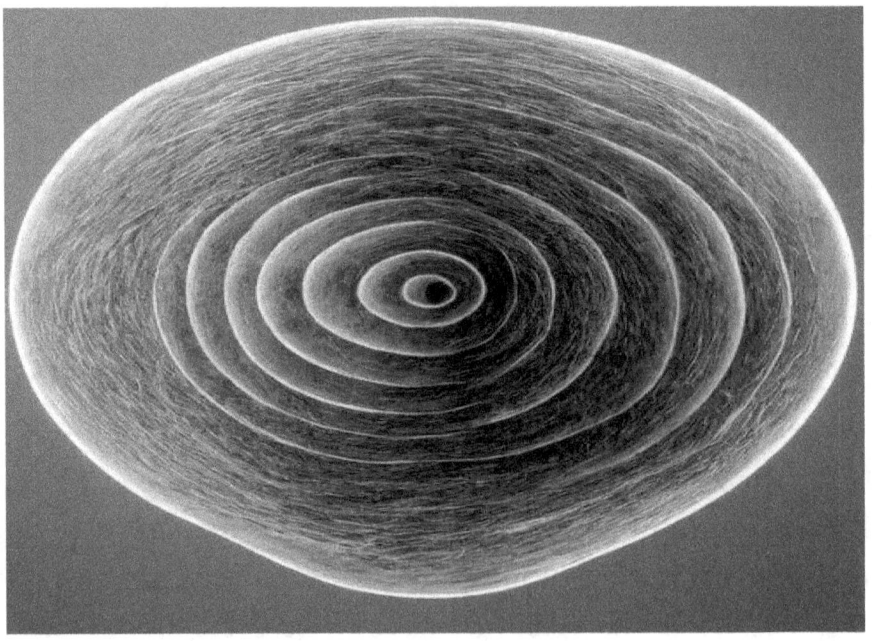

In the aftermath of a seismic event, when the dust settles and the ground stills, it is our resilience that will guide us forward. Resilience is the ability to adapt, to learn, and to grow from adversity. It is the courage to face challenges head-on and emerge stronger on the other side.

As you close this book, let it serve as a reminder of the remarkable human capacity for resilience. Carry this knowledge with you, not as a burden, but as a source of empowerment. For in the face of nature's fury, it is our preparedness and resilience that will light the way towards a future where we not only survive but thrive.

Embrace resilience, inspire preparedness, and let this book be the catalyst for a lifelong commitment to being ready for the unexpected. Together, we can build a world where the tremors of the Earth are met with unwavering strength and a resolute spirit that cannot be shaken.

# THE END

www.ingramcontent.com/pod-product-compliance
Lightning Source LLC
Chambersburg PA
CBHW050028230526
45470CB00003B/1176